Farmers

Julie Murray

Abdo Kids Junior
is an Imprint of Abdo Kids
abdobooks.com

Abdo
MY COMMUNITY: JOBS
Kids

abdobooks.com

Published by Abdo Kids, a division of ABDO, P.O. Box 398166, Minneapolis, Minnesota 55439.

Abdo Kids Junior™ is a trademark and logo of Abdo Kids.

Printed in China

102020

012021

Photo Credits: iStock, Shutterstock

Production Contributors: Teddy Borth, Jennie Forsberg, Grace Hansen

Design Contributors: Candice Keimig, Dorothy Toth

Library of Congress Control Number: 2020910586

Publisher's Cataloging-in-Publication Data

Names: Murray, Julie, author.

Title: Farmers / by Julie Murray

Description: Minneapolis, Minnesota : Abdo Kids, 2021 | Series: My community: jobs | Includes online resources and index.

Identifiers: ISBN 9781098205812 (lib. bdg.) | ISBN 9781098206376 (ebook) | ISBN 9781098206659 (Read-to-Me ebook)

Subjects: LCSH: Farmers--Juvenile literature. | Agriculture--Juvenile literature. | Community life--Juvenile literature. | Occupations--Juvenile literature. | Cities and towns--Juvenile literature.

Classification: DDC 630.203--dc23

Table of Contents

Farmers

Some farmers **produce** food. Their food feeds people.

5

Some raise animals. The animals have to be cared for.

Hank has chickens.

He collects the eggs.

Joy milks the cows. The milk is used to make ice cream.

Sally feeds the pigs.

They eat a lot!

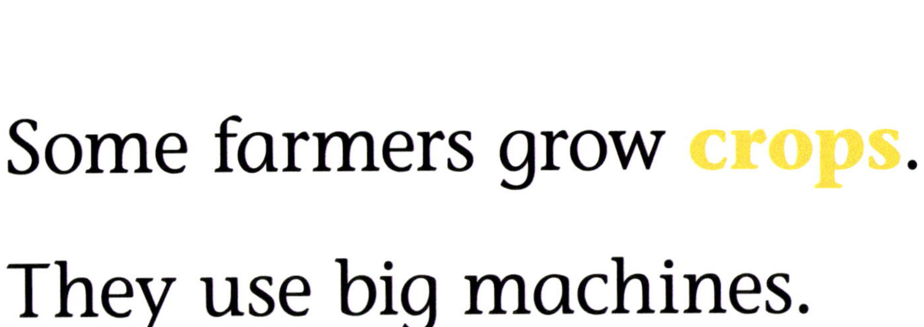

Some farmers grow **crops**.

They use big machines.

Jane grows corn. The stalks are tall!

The wheat is **harvested**. It will be used to make bread.

Sara grows radishes. She sells them at a farmer's market.

100%
ORGANIC

A Farmer's Tools

combine

plow

seeds

tractor

Glossary

crop
plants grown on a farm.

harvested
gathered.

produce
make or grow.

Index

Visit **abdokids.com** to access crafts, games, videos, and more!

Use Abdo Kids code

MFK5812

or scan this QR code!